Accident Source Terms for Light-Water Nuclear Power Plants

Final Report

U.S. Nuclear Regulatory Commission

Office of Nuclear Regulatory Research

L. Soffer, S. B. Burson, C. M. Ferrell,
R. Y. Lee, J. N. Ridgely

Accident Source Terms for Light-Water Nuclear Power Plants

Final Report

Manuscript Completed: February 1995
Date Published: February 1995

L. Soffer, S. B. Burson, C. M. Ferrell,
R. Y. Lee, J. N. Ridgely

Division of Systems Technology
Office of Nuclear Regulatory Research
U.S. Nuclear Regulatory Commission
Washington, DC 20555–0001

Abstract

In 1962 The U.S. Atomic Energy Commission published TID-14844, "Calculation of Distance Factors for Power and Test Reactors" which specified a release of fission products from the core to the reactor containment in the event of a postulated accident involving a "substantial meltdown of the core." This "source term," the basis for the NRC's Regulatory guides 1.3 and 1.4, has been used to determine compliance with the NRC's reactor site criteria, 10 CFR Part 100, and to evaluate other important plant performance requirements. During the past 30 years substantial additional information on fission product releases has been developed based on significant severe accident research. This document utilizes this research by providing more realistic estimates of the "source term" release into containment, in terms of timing, nuclide types, quantities, and chemical form, given a severe core-melt accident. This revised "source term" is to be applied to the design of future Light Water Reactors (LWRs). Current LWR licensees may voluntarily propose applications based upon it. These will be reviewed by the NRC staff.

CONTENTS

TABLES

CONTENTS (Cont'd)

APPENDICES

Preface

In 1962, the Atomic Energy Commission issued Technical Information Document (TID) 14844, "Calculation of Distance Factors for Power and Test Reactors." In this document, a release of fission products from the core of a light-water reactor (LWR) into the containment atmosphere ("source term") was postulated for the purpose of calculating off-site doses in accordance with 10 CFR Part 100, "Reactor Site Criteria." The source term postulated an accident that resulted in substantial meltdown of the core, and the fission products assumed released into the containment were based on an understanding at that time of fission product behavior. In addition to site suitability, the regulatory applications of this source term (in conjunction with the dose calculation methodology) affect the design of a wide range of plant systems.

In the past 30 years, substantial information has been developed updating our knowledge about severe LWR accidents and the resulting behavior of the released fission products. The purpose of this document is to provide a postulated fission product source term released into containment that is based on current understanding of LWR accidents and fission product behavior. The information contained in this document is applicable to LWR designs and is intended to form the basis for the development of regulatory guidance, primarily for future LWRs. This report will serve as a basis for possible changes to regulatory requirements. However, acceptance of any proposed changes will be on a case-by-case basis.

Source terms for future reactors may differ from those presented in this report which are based upon insights derived from current generation light-water reactors. An applicant may propose changes in source term parameters (timing, release magnitude, and chemical form) from those contained in this report, based upon and justified by design specific features.

1 INTRODUCTION AND BACKGROUND

1.1 Regulatory Use of Source Terms

The use of postulated accidental releases of radioactive materials is deeply embedded in the regulatory policy and practices of the U.S. Nuclear Regulatory Commission (NRC). For over 30 years, the NRC's reactor site criteria in 10 CFR Part 100 (Ref. 1) have required, for licensing purposes, that an accidental fission product release resulting from "substantial meltdown" of the core into the containment be postulated to occur and that its potential radiological consequences be evaluated assuming that the containment remains intact but leaks at its maximum allowable leak rate. Radioactive material escaping from the containment is often referred to as the "radiological release to the environment." The radiological release is obtained from the containment leak rate and a knowledge of the airborne radioactive inventory in the containment atmosphere. The radioactive inventory within containment is referred to as the "in-containment accident source term."

The expression "in-containment accident source term," as used in this document, denotes the radioactive material composition and magnitude, as well as the chemical and physical properties of the material within the containment that are available for leakage from the reactor to the environment. The "in-containment accident source term" will normally be a function of time and will involve consideration of fission products being released from the core into the containment as well as removal of fission products by plant features intended to do so (e.g., spray systems) or by natural removal processes.

For currently licensed plants, the characteristics of the fission product release from the core into the containment are set forth in Regulatory Guides 1.3 and 1.4 (Refs. 2,3) and have been derived from the 1962 report, TID-14844 (Ref. 4). This release consists of 100% of the core inventory of noble gases and 50% of the iodines (half of which are assumed to deposit on interior surfaces very rapidly). These values were based largely on experiments performed in the late 1950s involving heated irradiated UO_2 pellets. TID-14844 also included 1% of the remaining solid fission products, but these were dropped from consideration in Regulatory Guides 1.3 and 1.4. The 1% of the solid fission products are considered in certain areas such as equipment qualification.

Regulatory Guides 1.3 and 1.4 (Refs. 2 and 3) specify that the source term within containment is assumed to be instantaneously available for release and that the iodine chemical form is assumed to be predominantly (91%) in elemental (I_2) form, with 5% assumed to be particulate iodine and 4% assumed to be in organic form. These assumptions have significantly affected the design of engineered safety features. Containment isolation valve closure times have also been affected by these assumptions.

Use of the TID-14844 release has not been confined to an evaluation of site suitability and plant mitigation features such as sprays and filtration systems. The regulatory applications of this release are wide, including the basis for (1) the post-accident radiation environment for which safety-related equipment should be qualified, (2) post-accident habitability requirements for the control room, and (3) post-accident sampling systems and accessibility.

In contrast to the TID-14844 source term and containment leakage release used for design basis accidents, severe accident releases to the environment first arose in probabilistic risk assessments (e.g., Reactor Safety Study, WASH-1400 (Ref. 5)) in examining accident sequences that involved core melt and containments that could fail. Severe accident releases represent mechanistically determined best estimate releases to the environment, including estimates of failures of containment integrity. This is very different from the combination of the non-mechanistic release to containment postulated by TID-14844 coupled with the assumption of very limited containment leakage used for Part 100 siting calculations for design basis accidents. The worst severe accident releases resulting from containment failure or containment bypass can lead to consequences that are much greater than those associated with a TID-14844 source term released into containment where the containment is assumed to be leaking at its maximum leak rate for its design conditions. Indeed, some of the most severe releases arise from some containment bypass events, such as rupture of multiple steam generator tubes.

Although severe accident source terms have not been used in individual plant licensing safety evaluations, they have had significant regulatory applications. Source terms from severe accidents (beyond-design-basis accidents) came into regulatory consideration and usage shortly after the issuance of WASH-1400 in 1975, and their application was accelerated after the Three Mile Island accident in March 1979. Current applications rely to a large extent on the results of WASH-1400 and include (1) part of the basis for the sizes of emergency planning zones for all plants, (2) the basis for staff assessments of severe accident risk in plant environmental impact statements, and (3) part of the basis for staff prioritization and resolution of generic safety issues, unresolved safety issues, and other regulatory analyses. Source term assessments based on WASH-1400 methodology appear in many probabilistic risk assessment studies performed to date.

1.2 Research Insights Since TID-14844

Source term estimates under severe accident conditions became of great interest shortly after the Three-Mile Island (TMI) accident when it was observed that only relatively small amounts of iodine were released to the environment compared with the amount predicted to be released in licensing calculations. This led a number of observers to claim that severe accident releases were much lower than previously estimated.

The NRC began a major research effort about 1981 to obtain a better understanding of fission-product transport and release mechanisms in LWRs under severe accident conditions. This research effort has included extensive NRC staff and contractor efforts involving a number of national laboratories as well as nuclear industry groups. These cooperative research activities resulted in the development and application of a group of computer codes known as the Source Term Code Package (STCP) (Ref. 6) to examine core-melt progression and fission product release and transport in LWRs. The NRC staff has also sponsored significant review efforts by peer reviewers, foreign partners in NRC research programs, industry groups, and the general public. The STCP methodology for severe accident source terms has also been reflected in NUREG-1150 (Ref. 7), which provides an updated risk assessment for five U.S. nuclear power plants.

As a result of the NRC's research effort to obtain a better understanding of fission product transport and release mechanisms in LWRs under severe accident conditions, the STCP emerged as an integral tool for analysis of fission product transport in the reactor coolant system (RCS) and containment. The STCP models release from the fuel with CORSOR (Ref. 8) and fission product retention and transport in the RCS with TRAPMELT (Ref. 9). Releases from core-concrete interactions are modeled using the VANESA and CORCON (Ref. 10) codes. Depending upon the containment type, SPARC or ICEDF (Refs. 11,12) are used in conjunction with NAUA (Ref.13) to model the transport and retention of fission product releases from the RCS and from core-concrete interactions into the containment, with subsequent release of fission products to the environment consistent with the state of the containment.

Improved modeling of severe accident phenomena, including fission product transport, has been provided by the recently developed MELCOR (Ref. 14) code. At this time, however, an insufficient body of calculations is available to provide detailed insights from this model.

Using analyses based on the STCP and MELCOR codes and NUREG-1150, the NRC has sponsored studies (Refs. 15–17) that analyzed the timing, magnitude, and duration of fission product releases. In addition, an examination and assessment of the chemical form of iodine likely to be found within containment as a result of a severe accident has also been carried out (Ref. 18).

In contrast to the instantaneous releases that were postulated in Regulatory Guides 1.3 and 1.4, analyses of severe accident sequences have shown that, despite differences in plant design and accident sequence, such releases can be generally categorized in terms of phenomenological phases associated with the degree of fuel melting and relocation, reactor pressure vessel integrity, and, as applicable, attack upon concrete below the reactor cavity by molten core materials. The general phases, or progression, of a severe LWR accident are shown in Table 1.1.

Table 1.1 Release Phases of a Severe Accident

Release Phases
Coolant Activity Release
Gap Activity Release
Early In-Vessel Release
Ex-Vessel Release
Late In-Vessel Release

Initially there is a release of coolant activity associated with a break or leak in the reactor coolant system. Assuming that the coolant loss cannot be accommodated by the reactor coolant makeup systems or the emergency core cooling systems, fuel cladding failure would occur with a release of the activity located in the gap between the fuel pellet and the fuel cladding.

As the accident progresses, fuel degradation begins, resulting in a loss of fuel geometry accompanied by gradual melting and slumping of core materials to the bottom of the reactor pressure vessel. During this period, the early in-vessel release phase, virtually all the noble gases and significant fractions of the volatile nuclides such as iodine and cesium are released into containment. The amounts of volatile nuclides released into containment during the early in-vessel phase are strongly influenced by the residence time of the radioactive material within the RCS during core degradation. High pressure sequences result in long residence times and significant retention and plateout of volatile nuclides within the RCS, while low pressure sequences result in relatively short residence times and little retention within the RCS and consequently higher releases into containment.

If failure of the bottom head of the reactor pressure vessel occurs, two additional release phases may occur. Molten core debris released from the reactor pressure vessel into the containment will interact with the concrete structural materials of the cavity below the reactor (ex-vessel release phase). As a result of these interactions, quantities of the less volatile nuclides may be released into containment. Ex-vessel releases are influenced somewhat by the type of concrete in the reactor cavity. Limestone concrete decomposes to produce greater quantities of CO and CO_2 gases than basaltic concrete. These gases may, in turn, sparge some of the less volatile nuclides, such as barium and strontium, and small fractions of the lanthanides into the containment atmosphere. Large quantities of non-radioactive aerosols may also be released as a result of core-concrete interactions. The presence of water in the reactor cavity overlying any core debris can significantly reduce the ex-vessel releases (both radioactive and non-radioactive) into the containment, either by cooling the core debris, or at least by scrubbing the releases and retaining a large fraction in the water. The degree of scrubbing will depend, of course, upon the depth and temperature of any water overlying the core debris. Simultaneously, and generally with a longer duration, late in-vessel releases of some of the volatile nuclides, which had deposited in the reactor coolant system during the in-vessel phase, will also occur and be released into containment.

Two other phenomena that affect the release of fission products into containment could also occur, as discussed in Reference 7. The first of these is referred to as "high pressure melt ejection" (HPME). If the RCS is at high pressure at the time of failure of the bottom head of the reactor pressure vessel, quantities of molten core materials could be injected into the containment at high velocities. In addition to a potentially rapid rise in containment temperature, a significant amount of radioactive material could also be added to the containment atmosphere, primarily in the form of aerosols. The occurrence of HPME is precluded at low RCS pressures. A second phenomenon that could affect the release of fission products into containment is a possible steam explosion as a result of interactions between molten core debris and water. This could lead to fine fragmentation of some portion of the molten core debris with an increase in the amount of airborne fission products. While small scale steam explosions are considered quite likely to occur, they will not result in significant increases in the airborne activity already within containment. Large scale steam explosions, on the other hand, could result in significant increases in airborne activity, but are much less likely to occur. In any event, releases of particulates or vapors during steam explosions will also be accompanied by large amounts of water droplets, which would tend to quickly sweep released material from the atmosphere.

2 OBJECTIVES AND SCOPE

2.1 General

The primary objective of this report is to define a revised accident source term for regulatory application for future LWRs. The intent is to capture the major relevant insights available from recent severe accident research on the phenomenology of fission product release and transport behavior. The revised source term is expressed in terms of times and rates of appearance of radioactive fission products into the containment, the types and quantities of the species released, and other important attributes such as the chemical forms of iodine. This mechanistic approach will therefore present, for regulatory purposes, a more realistic portrayal of the amount of fission products present in the containment from a postulated severe accident.

2.2 Accidents Considered

In order to determine accident source terms for regulatory purposes, a range of severe accidents that have been analyzed for LWR plants was examined. Evaluation of a range of severe accident sequences was based upon work done in support of NUREG-1150 (Ref. 7). This work is documented in NUREG/CR-5747 (Ref. 17) and employed the integrated Source Term Code Package (STCP) computer codes, together with insights from the MELCOR code, which were used to analyze specific accident sequences of interest to provide the accident chronology as well as detailed estimates of fission product behavior within the reactor coolant system and the other pertinent parts of the plant. The sequences studied progressed to a complete core melt, involving failure of the reactor pressure vessel and including core-concrete interactions, as well.

A key decision to be made in defining an accident source term is the severity of the accident or group of accidents to be considered. Footnote 1 to 10 CFR Part 100 (Ref. 1), in referring to the postulated fission product release to be used for evaluating sites, notes that "Such accidents have generally been assumed to result in substantial meltdown of the core with subsequent release of appreciable quantities of fission products." Possible choices range from (1) slight fuel damage accidents involving releases into containment of a small fraction of the volatile nuclides such as the noble gases, (2) severe core damage accidents involving major fuel damage but without reactor vessel failure or core-concrete interactions (similar in severity to the TMI accident), or (3) complete core-melt events with core-concrete interactions. These outcomes are not equally probable. Since many reactor systems must fail for core degradation with reactor vessel failure to occur and core-concrete interactions to occur, one or more systems may be returned to an operable status before core melt commences. Hence, past operational and accident experience together with information on modern plant designs, together with a vigorous program aimed at developing accident management procedures, indicate that complete core-melt events resulting in reactor pressure vessel failure are considerably less likely to occur than those involving major fuel damage without reactor pressure vessel failure, and that these, in turn, are less likely to occur than those involving slight fuel damage.

For completeness, this report displays the mean or average release fractions for all the release phases associated with a complete core melt. However, it is concluded that any source term selected for a particular regulatory application should appropriately reflect the likelihood associated with its occurrence.

It is important to emphasize that the release fractions for the source terms presented in this report are intended to be representative or typical, rather than conservative or bounding values, of those associated with a low pressure core-melt accident, except for the initial appearance of fission products from failed fuel, which was chosen conservatively. The release fractions are not intended to envelope all potential severe accident sequences, nor to represent any single sequence, since accident sequences yielding both higher as well as lower release fractions were examined and factored into the final report presented here.

Source terms for future reactors may differ from those presented in this report which are based upon insights derived from current generation light-water reactors. An applicant may propose changes in source term parameters (timing, release magnitude, and chemical form) from those contained in this report, based upon and justified by design specific features.

The NRC staff also intends to allow credit for removal or reduction of fission products within containment via engineered features provided for fission product reduction such as sprays or filters, as well as by natural processes such as aerosol deposition. These are discussed in Section 5.

2.3 Limitations

The accident source terms defined in this report have been derived from examination of a set of severe accident sequences for LWRs of current design. Because of general similarities in plant and core design parameters, these results are also considered to be applicable to evolutionary LWR designs such as General Electric's Advanced Boiling Water Reactor (ABWR) and Combustion Engineering's (CE) System 80+.

Currently, the NRC staff is reviewing reactor designs for several smaller LWRs employing some passive features for core cooling and containment heat removal. While the "passive" plants are generally similar to present LWRs, they are expected to have somewhat lower core power densities than those of current LWRs. Hence, an accident for the passive plants similar to those used in this study would likely extend over a longer time span. For this reason, the timing and duration values provided in the release tables given in Section 3.3 are probably shorter than those applicable to the passive plants. The release fractions shown may also be overestimated somewhat for high pressure sequences associated with the passive plants, since longer times for accident progression would also allow for enhanced retention of fission products in the primary coolant system during core heatup and degradation. Despite the lack of specific accident sequence information for these designs, the in-containment accident source terms provided below may be considered generally applicable to the "passive" designs.

The accident source terms provided in this report are not considered applicable to reactor designs that are very different from LWRs, such as high-temperature gas-cooled reactors or liquid-metal reactors.

Recent information has indicated that high burnup fuel, that is, fuel irradiated at levels in excess of about 40 GWD/MTU, may be more prone to failure during design basis reactivity insertion accidents (RIA) than previously thought. Preliminary indications are that high burnup fuel also may be in a highly fragmented or powdered form, so that failure of the cladding could result in a significant fraction of the fuel itself being released. In contrast, the source term contained in this report is based upon fuel behavior results obtained at lower burnup levels where the fuel pellet remains intact upon cladding failure, resulting in a release only of those fission product gases residing in the gap between the fuel pellet and the cladding. Because of this recent information regarding high burnup fuels, the NRC staff cautions that, until further information indicates otherwise, the source term in this report (particularly gap activity) may not be applicable for fuel irradiated to high burnup levels (in excess of about 40 GWD/MTU).

3 ACCIDENT SOURCE TERMS

The expression "in-containment source terms," as used in this report, denotes the fission product inventory present in the containment at any given time during an accident. To evaluate the in-containment source term during the course of an accident, the time-history of the fission product release from the core into the containment must be known, as well as the effect of fission product removal mechanisms, both natural and engineered, to remove radioactive materials from the containment atmosphere. This section discusses the time-history of the fission product releases into the containment. Removal mechanisms are discussed in Section 5.

3.1 Accident Sequences Reviewed

All the accident sequences identified in NUREG-1150 were reviewed and some additional Source Term Code Package (STCP) and MELCOR calculations were performed. The dominant sequences which are

considered to significantly impact the source term are summarized in Table 3.1 for BWRs and Table 3.2 for PWRs.

3.2 Onset of Fission Product Release

This section discusses the assumptions used in selecting the scenario appropriate for defining the early phases of the source term (coolant activity and gap release phases). It was considered appropriate to base these early release phases on the design basis initiation that could lead to earliest fuel failures.

A review of current plant final safety analysis reports (FSARs) was made to identify all design basis accidents in which the licensee had identified fuel failure. For all accidents with the potential for release of radioactivity into the environment, the class of accident that had the shortest time until the first fuel rod failed was the design basis LOCA. As might be expected, the time until cladding failure is very sensitive to the design of the reactor, the type of accident assumed, and the fuel rod design. In particular, the maximum linear heat generation rate, the internal fuel rod pressure, and the stored energy in the fuel rod are significant considerations.

Table 3.1 BWR Source Term Contributing Sequences

Plant	Sequence	Description
Peach Bottom	TC1	ATWS with reactor depressurized
	TC2	ATWS with reactor pressurized
	TC3	TC2 with wetwell venting
	TB1	SBO with battery depletion
	TB2	TB1 with containment failure at vessel failure
	S2E1	LOCA (2"), no ECCS and no ADS
	S2E2	S2E1 with basaltic concrete
	V	RHR pipe failure outside containment
	TBUX	SBO with loss of all DC power
LaSalle	TB	SBO with late containment failure
Grand Gulf	TC	ATWS early containment failure fails ECCS
	TB1	SBO with battery depletion
	TB2	TB1 with H_2 burn fails containment
	TBS	SBO, no ECCS but reactor depressurized
	TBR	TBS with AC recovery after vessel failure

SBO	Station Blackout	LOCA	Loss of Coolant Accident
RCP	Reactor Coolant Pump	RHR	Residual Heat Removal
ADS	Automatic Depressurization System	ATWS	Anticipated Transient Without Scram

Table 3.2 PWR Source Term Contributing Sequences

Plant	Sequence	Description
Surry	AG	LOCA (hot leg), no containment heat removal systems
	TMLB'	LOOP, no PCS and no AFWS
	V	Interfacing system LOCA
	S3B	SBO with RCP seal LOCA
	S2D-δ	SBLOCA, no ECCS and H_2 combustion
	S2D-β	SBLOCA with 6" hole in containment
Zion	S2DCR	LOCA (2"), no ECCS no CSRS
	S2DCF1	LOCA RCP seal, no ECCS, no containment sprays, no coolers—H_2 burn or DCH fails containment
	S2DCF2	S2DCF1 except late H_2 or overpressure failure of containment
	TMLU	Transient, no PCS, no ECCS, no AFWS—DCH fails containment
Oconee 3	TMLB'	SBO, no active ESF systems
	S1DCF	LOCA (3"), no ESF systems
Sequoyah	S3HF1	LOCA RCP, no ECCS, no CSRS with reactor cavity flooded
	S3HF2	S3HF1 with hot leg induced LOCA
	3HF3	S3HF1 with dry reactor cavity
	S3B	LOCA (½") with SBO
	TBA	SBO induces hot leg LOCA—hydrogen burn fails containment
	ACD	LOCA (hot leg), no ECCS no CS
	S3B1	SBO delayed 4 RCP seal failures, only steam driven AFW operates
	S3HF	LOCA (RCP seal), no ECCS, no CSRS
	S3H	LOCA (RCP seal) no ECC recirculation

SBO	Station Blackout	LOCA	Loss of Coolant Accident
RCP	Reactor Coolant Pump	DCH	Direct Containment Heating
PCS	Power Conversion System	ESF	Engineered Safety Feature
CS	Containment Spray	CSRS	CS Recirculation System
ATWS	Anticipated Transient Without Scram	LOOP	Loss of Offsite Power

The details of the specific accident sequences are documented in NUREG/CR-5747, Estimate of Radionuclide Release Characteristics into Containment Under Severe Accident Conditions (Ref. 17).

To determine whether a design basis LOCA was a reasonable scenario upon which to base the timing of initial fission product release into the containment, various PRAs were reviewed to determine the contribution to core damage frequency (CDF) resulting from LOCAs. This information is shown in Table 3.3. As can be seen from this table, LOCAs are a small contributor to CDF for BWRs, but can be a substantial contributor for PWRs. Therefore, for PWRs a large LOCA is considered a reasonable initiator to assume for modeling the earliest appearance of the gap activity if the plant has not been approved for leak before break (LBB) operation. For plants that have received LBB approval, a small LOCA (6" line break) would more appropriately model the timing. For BWRs, large LOCAs may not be an appropriate scenario for gap activity timing. However, since the time to initial fuel rod failure is long for BWRs, even for large LOCAs, use of the large LOCA scenario should not unduly penalize BWRs and will maintain consistency with the assumptions for the PWR. As with the PWR, for an LBB approved plant, the timing associated with a small LOCA (6" line break) would be more appropriate.

In order to provide a realistic estimate of the shortest time for fuel rod failure for the LOCA, calculations were performed using the FRAPCON2, SCDAP/RELAP5 MOD 3.0, and FRAPT6 computer codes for two plants. The two plants were a Babcock and Wilcox (B&W) plant with a 15 by 15 fuel rod array and a Westinghouse 4-loop (W) plant with a 17 by 17 fuel rod array. For each plant, a sensitivity study was

performed to identify the size of the LOCA that resulted in the shortest fuel rod failure time (Ref. 15). In both cases, the accident was a double-ended guillotine rupture of the cold leg pipe. The minimum time from the time of accident initiation until first fuel rod failure was calculated to be 13 and 24.6 seconds for the B&W and W plants, respectively. A sensitivity study was performed to determine the effect of tripping or not tripping the reactor coolant pumps. The results indicated that tripping of the reactor coolant pumps had no appreciable impact on timing. For a 6-inch line break, the time until first fuel rod failure is expected to be greater than 6.5 and 10 minutes, respectively.

Table 3.3 Contribution of LOCAs to Core Damage Frequency (CDF)—Internal Events

Boiling Water Reactors	Percent of CDF caused by LOCAs	Percent of CDF caused by large LOCAs (>6" line break)
Peach Bottom (NUREG-1150)	3.5	1.0
Grand Gulf (NUREG-1150)	0.1	0.03
Millstone 1 (Utility)	23	13
Pressurized Water Reactors		
Surry (NUREG-1150)	15	4.3
Sequoyah (NUREG-1150)	63	4.6
Zion (NUREG-1150)	87	1.4
Calvert Cliffs (IREP)	21	<1
Oconee-3 (EPRI/NSAC)	43	3.0

A comparison calculation was done using the TRAC-PF1 MOD 1 code, version 14.3U5Q.LG on the W plant. This analysis indicated that the first fuel rod failure would occur 34.9 seconds after pipe rupture, in contrast to the value of 24.6 seconds calculated using SCDAP/RELAP. The reasons for the difference between the SCDAP/RELAP5 MOD 3.0 and TRAC-PF1 MOD 1 are discussed in Reference 15.

The review of the FSARs for BWRs indicates that fuel failures may occur significantly later, on the order of several minutes or more. No calculations have been performed using the aforementioned suite of codes.

For determining the time of appearance of gap activity in the containment (i.e., initial fuel failure), which corresponds to the duration of the coolant activity phase and the beginning of the gap activity phase, it would be appropriate to perform a plant specific calculation using the codes described above. However, if no plant specific calculations are performed, the minimum times discussed above may be used to provide an estimate of the earliest time to fuel rod failure.

Source terms for future reactors may differ from those presented in this report which are based upon insights derived from current generation light-water reactors. An applicant may propose changes in source term parameters (timing, release magnitude, and chemical form) from those contained in this report, based upon and justified by design specific features.

3.3 Duration of Release Phases

Section 1.2 provided a qualitative discussion of the release phases of an accident. This section provides estimated durations for these release phases.

The coolant activity phase begins with a postulated pipe rupture and ends when the first fuel rod has been estimated to fail. During this phase, the activity released to the containment atmosphere is that associated with very small amounts of radioactivity dissolved in the coolant itself. As discussed in Section 3.2 above, this phase is estimated to last about 25 seconds for Westinghouse PWRs, and about 13 seconds for B&W PWRs, assuming a large break LOCA. For a smaller LOCA (e.g., a 6-inch line break), such as would be considered for a plant that has received LBB approval, the coolant activity phase duration would be expected to be at least 10 minutes. Although not specifically evaluated at this time, Combustion Engineering (CE) PWRs would be expected to have coolant activity durations similar to Westinghouse plants. For BWRs, the coolant activity phase would be expected to last longer; however, unless plant specific calculations are made, the durations discussed above are considered applicable.

The gap activity release phase begins when fuel cladding failure commences. This phase involves the release of that radioactivity that has collected in the gap between the fuel pellet and cladding. This process releases to containment a few percent of the total inventory of the more volatile radionuclides, particularly noble gases, iodine, and cesium. During this phase, the bulk of the fission products continue to be retained in the fuel itself. The gap activity phase ends when the fuel pellet bulk temperature has been raised sufficiently that significant amounts of fission products can no longer be retained in the fuel. As noted in Reference 16, a review of STCP calculated results for six reference plants, PWRs as well as BWRs, indicated that significant fission product releases from the bulk of the fuel itself were estimated to commence no earlier than about 30 minutes and 60 minutes for PWRs and BWRs, respectively, after the onset of the accident. However, more recent calculations (Ref. 19) for the Peach Bottom plant using the MELCOR code indicated that the durations of the gap release for three BWR accident sequences were about 30 minutes, as well. On this basis, the duration of the gap activity release phase has been selected to be 0.5 hours, for both PWRs and BWRs.

During the early in-vessel release phase, the fuel as well as other structural materials in the core reach sufficiently high temperatures that the reactor core geometry is no longer maintained and fuel and other materials melt and relocate to the bottom of the reactor pressure vessel. During this phase, significant quantities of the volatile nuclides in the core inventory as well as small fractions of the less volatile nuclides are estimated to be released into containment. This release phase ends when the bottom head of the reactor pressure vessel fails, allowing molten core debris to fall onto the concrete below the reactor pressure vessel. Release durations for this phase vary depending on both the reactor type and the accident sequence. Tables 3.4 and 3.5, based on results from Reference 16, show the estimated duration times for PWRs and BWRs, respectively.

Table 3.4 In-Vessel Release Duration for PWR Sequences

Plant	Accident Sequence*		Release Duration (Min)
Surry	TMLB'	(H)	41
Surry	S3B	(H)	36
Surry	AG	(L)	215
Surry	V	(L)	104
Zion	TMLU	(H)	41
Zion	S2DCR/S2DCF	(H)	39
Sequoyah	S3HF/S3B	(H)	46
Sequoyah	S3B1	(H)	75
Sequoyah	TMLB'	(H)	37
Sequoyah	TBA	(L)	195
Sequoyah	ACD	(L)	73
Oconee	TMLB'	(H)	35
Oconee	S1DCF	(L)	84

*(H or L) Denotes whether the accident occurs at high or low pressure.

Based on the information in these tables, the staff concludes that the in-vessel release phase is somewhat longer for BWR plants than for PWR plants. This is largely due to the lower core power density in BWR plants that extends the time for complete core melt. Representative times for the duration of the in-vessel release phase have been selected to be 1.3 hours and 1.5 hours, for PWR and BWR plants respectively, as recommended by Reference 17.

The ex-vessel release phase begins when molten core debris exits the reactor pressure vessel and ends when

Table 3.5 In-Vessel Release Duration for BWR Sequences

Plant	Accident Sequence*		Release Duration (Min)
Peach Bottom	TC2	(H)	66
Peach Bottom	TC3		68
Peach Bottom	TC1	(L)	97
Peach Bottom	TB1/TB2	(H)	91
Peach Bottom	V	(L)	69
Peach Bottom	S2E	(H)	81
Peach Bottom	TBUX	(H)	67
LaSalle	TB	(H)	81
Grand Gulf	TB	(H)	122
Grand Gulf	TC1	(L)	130
Grand Gulf	TBS/TBR	(L)	96

*(H or L) denotes whether the accident occurs at high or low pressure.

the debris has cooled sufficiently that significant quantities of fission products are no longer being released. During this phase, significant quantities of the volatile radionuclides not already released during the early in-vessel phase as well as lesser quantities of non-volatile radionuclides are released into containment. Although releases from core-concrete interactions are predicted to take place over a number of hours after vessel breach, Reference 16 indicates that the bulk of the fission products (about 90%), with the exception of tellurium and ruthenium, are expected to be released over a 2-hour period for PWRs and a 3-hour period for BWRs. For tellurium and ruthenium, ex-vessel releases extend over 5 and 6 hours, respectively, for PWRs and BWRs. The difference in duration of the ex-vessel phase between PWRs and BWRs is largely attributable to the larger amount of zirconium in BWRs, which provides additional chemical energy of oxidation. Based on Reference 17, the ex-vessel release phase duration is taken to be 2 and 3 hours, respectively, for PWRs and BWRs.

The late in-vessel release phase commences at vessel breach and proceeds simultaneously with the occurrence of the ex-vessel phase. However, the duration is not the same for both phases. During this release phase, some of the volatile nuclides deposited within the reactor coolant system earlier during core degradation and melting may re-volatilize and be released into containment. Reference 17, after a review of the source term uncertainty methodology used in NUREG-1150 (Ref. 7), estimates the late in-vessel

release phase to have a duration of 10 hours. This value has been selected for this report.

A summary of the release phases and the selected duration times for PWRs and BWRs is shown for reference purposes in Table 3.6.

Table 3.6
Release Phase Durations for PWRs and BWRs

Release Phase	Duration, PWRs (Hours)	Duration, BWRs (Hours)
Coolant Activity	10 to 30 seconds*	30 seconds*
Gap Activity	0.5	0.5
Early In-Vessel	1.3	1.5
Ex-Vessel	2	3
Late In-Vessel	10	10

*Without approval for leak-before-break. Coolant activity phase duration is assumed to be 10 minutes with leak-before-break approval.

3.4 Fission Product Composition and Magnitude

In considering severe accidents in which the containment might fail, WASH-1400 (Ref. 5) examined the spectrum of fission products and grouped 54 radionuclides into 7 major groups on the basis of similarity in chemical behavior. The effort associated with the STCP further analyzed these groupings and expanded the 7 fission product groups into 9 groups. These are shown in Table 3.7.

Table 3.7 STCP Radionuclide Groups

Group	Elements
1	Xe, Kr
2	I, Br
3	Cs, Rb
4	Te, Sb, Se
5	Sr
6	Ru, Rh, Pd, Mo, Tc
7	La, Zr, Nd, Eu, Nb, Pm, Pr, Sm, Y
8	Ce, Pu, Np
9	Ba

Both the results of the STCP analyses and the uncertainty analysis (using the results of the NUREG-1150 source term expert panel elicitation) reported in NUREG/CR-5747 (Ref. 17) indicate only minor differences between Ba and Sr releases. Hence, a revised grouping of radionuclides has been developed that groups Ba and Sr together. The relative importance to offsite health and economic consequences of the radioactive elements in a nuclear reactor core has been examined and documented in NUREG/CR-4467 (Ref. 20). In addition to the elements already included in Table 3.7, Reference 20 found that other elements such as Curium could be important for radiological consequences if released in sufficiently large quantities. For this reason, group 7 has been revised to include Curium (Cm) and Americium (Am), while group 6 has been revised to include Cobalt (Co). The revised radionuclide groups used in this report including revised titles and the elements comprising each group are shown in Table 3.8.

Source term releases into the containment were evaluated by reactor type, i.e., BWR or PWR, from the sequences in NUREG-1150 and the supplemental STCP calculations discussed in Section 3.1.

Releases into containment during the early in-vessel phase, prior to reactor pressure vessel failure, are markedly affected by retention in the RCS, which is a function of the residence time in the RCS during core degradation. High pressure in the RCS during core degradation allows for longer residence time of aerosols released from the core. This, in turn, permits increased retention of aerosols within the RCS and lower releases from the core into the containment.

Similarly, low pressure sequences cause aerosols generated within the RCS to be swept out rapidly without significant retention within the RCS, thereby resulting in higher release fractions from the core into containment.

Table 3.8 Revised Radionuclide Groups

Group	Title	Elements in Group
1	Noble gases	Xe, Kr
2	Halogens	I, Br
3	Alkali Metals	Cs, Rb
4	Tellurium group	Te, Sb, Se
5	Barium, strontium	Ba, Sr
6	Noble Metals	Ru, Rh, Pd, Mo, Tc, Co
7	Lanthanides	La, Zr, Nd, Eu, Nb, Pm, Pr, Sm, Y, Cm, Am
8	Cerium group	Ce, Pu, Np

The relative frequency of occurrence of high vs. low pressure sequences were examined for both BWRs and PWRs. The results of this survey are shown in Table 3.9, and they indicate that a significant fraction of the sequences examined, in terms of frequency, occurred at low pressure. In addition, advanced PWR designs are increasingly incorporating safety-grade depressurization systems, primarily to minimize the likelihood of high pressure melt ejection (HPME) with its associated high containment atmosphere heat loads and large amounts of atmospheric aerosols.

For these reasons, the composition and magnitude of the source term has been chosen to be representative of conditions associated with low pressure in the RCS at the time of reactor core degradation and pressure vessel failure. Reference 17 provides estimates of the mean core fractions released into containment, as estimated by NUREG-1150 (Ref. 7), for accident sequences occurring under low RCS pressure and high zirconium oxidation conditions. These are shown in Tables 3.10 and 3.11.

3.5 Chemical Form

The chemical form of iodine and its subsequent behavior after entering containment from the reactor coolant system have been documented in NUREG/CR-5732, Iodine Chemical Forms in LWR Severe Accidents (Ref. 18) and in ORNL/TM-12202, "Models of Iodine Behavior in Reactor Containments," (Ref. 21).

Table 3.9 Fraction of mean core damage frequency with high, intermediate, and low pressure sequences (internal events only unless otherwise noted)

Boiling Water Reactors	High Pressure at Vessel Breach	Intermed. press. at vessel breach	Low Pressure at Vessel Breach (<200 psi)	No vessel breach
LaSalle— external events only	0.27	N/A	0.67	0.06
LaSalle— internal events only	0.19	N/A	0.62	0.19
Grand Gulf	0.28	N/A	0.51	0.21
Peach Bottom	0.51	N/A	0.41	0.08
Pressurized Water Reactors				
Surry	0.06	0.07	0.37	0.50
Sequoyah	0.14	0.21	0.24	0.41
Zion	0.03	0.15	0.72	0.10

Table 3.10 Mean Values of Radionuclides Into Containment for BWRs, Low RCS Pressure, High Zirconium Oxidation

Nuclide	Early In-Vessel	Ex-Vessel	Late In-vessel
N. G.	1.0	0	0
I	0.27	0.37	0.07
Cs	0.2	0.45	0.03
Te	0.11	0.38	0.01
Sr	0.03	0.24	0
Ba	0.03	0.21	0
Ru	0.007	0.004	0
La	0.002	0.01	0
Ce	0.009	0.01	0

Table 3.11 Mean Values of Radionuclide Releases Into Containment for PWRs, Low RCS Pressure, High Zirconium Oxidation

Nuclide	Early In-Vessel	Ex-Vessel	Late In-vessel
N.G.	1.0	0	0
I	0.4	0.29	0.07
Cs	0.3	0.39	0.06
Te	0.15	0.29	0.025
Sr	0.03	0.12	0
Ba	0.04	0.1	0
Ru	0.008	0.004	0
La	0.002	0.015	0
Ce	0.01	0.02	0

The results from Ref. 18 indicate that iodine entering the containment is at least 95% CsI with the remaining 5% as I plus HI, with not less than 1% of each as I and HI. Once the iodine enters containment, however, additional reactions are likely to occur. In an aqueous environment, as expected for LWRs, iodine is expected to dissolve in water pools or plate out on wet surfaces in ionic form as I^-. Subsequently, iodine behavior within containment depends on the time and pH of the water solutions. Because of the presence of other dissolved fission products, radiolysis is expected to occur and lower the pH of the water pools. Without any pH control, the results indicate that large fractions of the dissolved iodine will be converted to elemental iodine and be released to the containment atmosphere. However, if the pH is controlled and maintained at a value of 7 or greater, very little (less than 1%) of the dissolved iodine will be converted to elemental iodine. Some considerations in achieving pH control are discussed in NUREG/CR-5950, "Iodine Evolution and pH Control," (Ref. 22).

Organic compounds of iodine, such as methyl iodide, CH_3I, can also be produced over time largely as a result of elemental iodine reactions with organic materials. Organic iodide formation as a result of reactor accidents has been surveyed in WASH-1233, "Review of Organic Iodide Formation Under Accident Conditions in Water-Cooled Reactors," (Ref. 23), and more recently in NUREG/CR-4327, "Organic Iodide Formation Following Nuclear Reactor Accidents," (Ref.24). From an analysis of a number of containment experiments, WASH-1233 concluded that, considering both non-radiolytic as well as radiolytic means, no more than 3.2 percent of the airborne iodine would be converted to organic iodides during the first two hours following a fission product release. The value of 3.2 percent was noted as a conservative upper limit and was judged to be considerably less, since it did not account, among other things, for decreased radiolytic formation of organic iodide due to iodine removal mechanisms within containment. Reference 24 also included results involving irradiated fuel elements, and concluded that the organic iodide concentration within containment would be about 1 percent of the iodine release concentration over a wide range of iodine concentrations.

A conversion of 4 percent of the elemental iodine to organic has been implicitly assumed by the NRC staff in Regulatory Guides 1.3 and 1.4, based upon an upper bound evaluation of the results in WASH-1233. However, in view of the results of Ref. 23 that a conversion of 3.2 percent is unduly conservative, a value of 3 percent is considered more realistic and will be used in this report. Where the pH is controlled at

values of 7 or greater within the containment, elemental iodine can be taken as comprising no more than 5 percent of the total iodine released, and iodine in organic form may be taken as comprising no greater than 0.15 percent (3 percent of 5 percent) of the total iodine released.

Organic iodide formation in BWRs versus PWRs is not notably different. Reference 18 examined not only iodine entering containment as CsI; but also considered other reactions that might lead to volatile forms of iodine within containment, such as reactions of CsOH with surfaces and revaporization of CsI from RCS surfaces. Reference 18 indicates (Table 2.4) that for the Peach Bottom TC2 sequence, the estimated percentage of iodine as HI was 3.2 percent, not notably less than the PWR sequences examined. While organic iodide is formed largely from reactions of elemental iodine, Ref. 22 clearly notes that reactions with HI may be important.

Although organic iodine is not readily removed by containment sprays or filter systems, it is unduly conservative to assume that organic iodine is not removed at all from the containment atmosphere, once generated, since such an assumption can result in an overestimate of long-term doses to the thyroid. References 23 and 24 discuss the radiolytic destruction of organic iodide, and Standard Review Plan Section (S.R.P.) 6.5.2 notes the above reference and indicates that removal of organic iodide may be considered on a case-by-case basis. A rational model for organic iodine behavior within containment would consider both its formation as well as destruction in a time-dependent fashion. Development of such a model, however, is beyond the scope of the present report.

Clearly, where the pH is not controlled to values of 7 or greater, significantly larger fractions of elemental iodine, as well as organic iodine may be expected within containment.

All other fission products, except for the noble gases and iodine, discussed above, are expected to be in particulate form.

3.6 Proposed Accident Source Terms

The proposed accident source terms, including their timing as well as duration, are listed in Tables 3.12 for BWRs and 3.13 for PWRs. The information for these tables was derived from the simplification of the NUREG-1150 (Ref. 7) source terms documented in NUREG/CR-5747 (Ref. 17). It should also be noted that the rate of release of fission products into the containment is assumed to be constant during the duration time shown.

Table 3.12 BWR Releases Into Containment*

	Gap Release***	Early In-Vessel	Ex-Vessel	Late In-Vessel
Duration (Hours)	0.5	1.5	3.0	10.0
Noble Gases**	0.05	0.95	0	0
Halogens	0.05	0.25	0.30	0.01
Alkali Metals	0.05	0.20	0.35	0.01
Tellurium group	0	0.05	0.25	0.005
Barium, Strontium	0	0.02	0.1	0
Noble Metals	0	0.0025	0.0025	0
Cerium group	0	0.0005	0.005	0
Lanthanides	0	0.0002	0.005	0

* Values shown are fractions of core inventory.
** See Table 3.8 for a listing of the elements in each group
*** Gap release is 3 percent if long-term fuel cooling is maintained.

Table 3.13 PWR Releases Into Containment*

	Gap Release***	Early In-Vessel	Ex-Vessel	Late In-Vessel
Duration (Hours)	0.5	1.3	2.0	10.0
Noble Gases**	0.05	0.95	0	0
Halogens	0.05	0.35	0.25	0.1
Alkali Metals	0.05	0.25	0.35	0.1
Tellurium group	0	0.05	0.25	0.005
Barium, Strontium	0	0.02	0.1	0
Noble Metals	0	0.0025	0.0025	0
Cerium group	0	0.0005	0.005	0
Lanthanides	0	0.0002	0.005	0

* Values shown are fractions of core inventory.
** See Table 3.8 for a listing of the elements in each group
*** Gap release is 3 percent if long-term fuel cooling is maintained.

It is emphasized that the release fractions for the source terms presented in this report are intended to be representative or typical, rather than conservative or bounding values, of those associated with a low pressure core-melt accident, except for the initial appearance of fission products from failed fuel, which was chosen conservatively. The release fractions are not intended to envelope all potential severe accident sequences, nor to represent any single sequence.

Tables 3.12 and 3.13 in this, the final report, were modified from the tables in the draft report which were taken from Table 3.9 and Table 3.10, for BWRs and

PWRs, respectively. The changes and the reasons for these was as follows:

1. BWR in-vessel release fractions for the volatile nuclides (I and Cs) increased slightly while ex-vessel release fractions for the same nuclides was reduced as a result of comments received and additional MELCOR calculations available after issuance of the draft report. The total I and Cs released into containment over all phases of the accident remained the same.

2. Release fractions for Te, Ba and Sr were reduced somewhat, both for in-vessel as well as ex-vessel releases, in response to comments.

3. Release fractions for the non-volatile nuclides, particularly during the early in-vessel phase were reduced significantly based on additional research results (Ref. 25) since issuance of NUREG-1150 which indicate that releases of low volatile nuclides, both in-vessel as well as ex-vessel, have been overestimated. A re-examination in response to comments received showed that the supposed "means" of the uncertainty distribution were in excess of other measures of the distribution, such as the 75th percentile. In this case, the 75th percentile was selected as an appropriate measure of the release fraction. For additional discussion on this topic, see Section 4.4.

4. Gap activity release fractions were reduced from 5 percent to 3 percent for accidents not involving degraded or molten core conditions, and where long-term fuel cooling is maintained. See additional discussion below.

Based on WASH-1400 (Ref. 5), the inventory of fission products residing in the gap between the fuel and the cladding is no greater than 3 percent except for cesium, which was estimated to be about 5 percent. NUREG/CR-4881 (Ref.16) reported a comparison of more recently available estimations and observations indicating that releases of the dominant fission product groups were generally below the values reported in Reference 5. However, the magnitude of fission products released during the gap release phase can vary, depending upon the type of accident. Accidents where fuel failures occur may be grouped as follows:

1. Accidents where long-term fuel cooling is maintained despite fuel failure. Examples include the design basis LOCA where ECCS functions, and a postulated spent fuel handling accident. For this category, fuel failure is taken to result in an immediate release, based upon References 5 and 16, of 3 percent of the volatile fission products (noble gases, iodine, and cesium) which are in the gap between the fuel pellet and the cladding. No subsequent appreciable release from the fuel pellet occurs, since the fuel does not experience prolonged high temperatures.

2. Accidents where long-term fuel cooling or core geometry are not maintained. Examples include degraded core or core-melt accidents, including the postulated limiting design basis fission product release into containment used to show compliance with 10 CFR Part 100. For this category, the gap release phase may overlap to some degree with the early in-vessel release phase. The release magnitude has been taken as an initial release of 3 percent of the volatiles (as for category 1), plus an additional release of 2 percent over the duration of the gap release phase.

3. Accidents where fuel failure results from reactivity insertion accidents (RIA), such as the postulated rod ejection (PWR) or rod drop (BWR) accidents. The accidents examined in this report do not contain information on reactivity induced accidents to permit a quantitative discussion of fission product releases from them. Hence, the gap release magnitude presented in Tables 3.12 and 3.13 may not be applicable to fission product releases resulting from reactivity insertion accidents.

Recent information has indicated that high burnup fuel, that is, fuel irradiated at levels in excess of about 40 GWD/MTU, may be more prone to failure during design basis reactivity insertion accidents than previously thought. Preliminary indications are that high burnup fuel also may be in a highly fragmented or powdered form, so that failure of the cladding could result in a significant fraction of the fuel itself being released. In contrast, the source term contained in this report is based upon fuel behavior results obtained at lower burnup levels where the fuel pellet remains intact upon cladding failure, resulting in a release only of those fission product gases residing in the gap between the fuel pellet and the cladding. Because of this recent information regarding high burnup fuels, the NRC staff cautions that, until further information indicates otherwise, the source term in Tables 3.12 and 3.13 (particularly gap activity) may not be applicable for fuel irradiated to high burnup levels (in excess of about 40 GWD/MTU).

With regard to the ex-vessel releases associated with core-concrete interactions, according to Reference 17, there were only slight differences in the fission products released into containment between limestone vs. basaltic concrete. Hence, the table shows the releases only for a limestone concrete. Further, the releases shown for the ex-vessel phase are assumed to be for a dry reactor cavity having no water overlying any core debris. Where water covers the core debris, aerosol scrubbing will take place and reduce the quantity of aerosols entering the containment atmosphere. See Section 5.4 for further information.

3.7 Nonradioactive Aerosols

In addition to the fission product releases into containment shown in Tables 3.12 and 3.13, quantities of nonradioactive or relatively low activity aerosols will also be released into containment. These aerosols arise from core structural and control rod materials released during the in-vessel phase and from concrete decomposition products during the ex-vessel phase. A detailed analysis of the quantity of nonfission product aerosols released into containment was not undertaken. Precise estimates of the masses of non-radioactive aerosols released into containment are difficult to determine.

Reference 26 evaluated one PWR sequence (Sequoyah) and one BWR (Peach Bottom) sequence and calculated in-vessel non-radioactive aerosol masses of 350 and 780 kilograms, respectively, for the PWR and BWR sequences. The same reference calculated that ex-vessel aerosol masses (assuming a dry cavity) would be higher, 3800 and 5600 kilograms, respectively, for the PWR and BWR sequences investigated. However, these values, particularly for the ex-vessel release phase, may be excessive. NUREG/CR–4624 (Ref. 27) examined several sequences for both PWRs and BWRs and calculated ex-vessel releases to containment of about 1000 and 4000 kilograms, respectively, for PWRs and BWRs. NUREG/CR–5942 (Ref.19), making use of the MELCOR code, calculated significantly lower releases during the ex-vessel phase of about 1000 kilograms for the Peach Bottom plant.

In view of the wide diversity of calculated results, the NRC staff concludes that precise estimates of the release of non-radioactive aerosols are not available at this time. Because nonradioactive aerosol masses could have an effect upon the operation of certain plant equipment, such as filter loadings or sump performance, during and following an accident, however, the NRC staff concludes that the release of non-radioactive aerosols should be considered by the designer using methods considered applicable for his design, and the potential impact upon the plant evaluated.

4 MARGINS AND UNCERTAINTIES

This section discusses some of the more significant conservatisms and margins in the proposed accident source term given in Section 3. Briefly, the proposed release fractions have been developed from a complete core-melt accident, that is, assuming core melt with reactor pressure vessel failure and with the assumption of core-concrete interactions. The timing aspects were selected to be typical of a low pressure core-melt scenario, except that the onset of the release of gap activity was based upon the earliest calculated time of fuel rod failure under accident conditions. The magnitude of the fission products released into containment was intended to be representative and, except for the low volatile nuclides, as discussed in section 4.4, was estimated from the mean values for a typical low-pressure core-melt scenario.

4.1 Accident Severity and Type

As noted earlier in Section 2.2, this report discusses mean or average release fractions for all the release phases associated with a complete core-melt accident, including reactor pressure vessel failure. The accident selected is one in which core melt occurs at low pressure conditions. A low pressure core melt scenario results in a relatively low level of fission product retention within the reactor coolant system, and a consequently high level of release of fission products from the core into containment during the early in-vessel release phase. Since the bulk of the fission products entering containment do so during the early in-vessel release phase, selection of a low pressure core melt scenario provides a high estimate of the total quantity of fission products released into containment, as well as that during the early in-vessel release phase.

4.2 Onset of Fission Product Release

The onset, or earliest time of appearance of fission products within containment, has been selected on the basis of the earliest time to failure of a fuel rod, given a design basis LOCA. This is estimated to be from about 13 to 25 seconds for plants that do not have leak-before-break approval for their reactor coolant system piping, and it is expected to vary depending on the reactor as well as the fuel rod design. This value, while representing some relaxation from the assumption of instantaneous appearance, is nevertheless conservative. As noted in Reference 15, these estimates are valid for a double-ended rupture of the largest pipe, assume that the fuel rod is being operated at the maximum peaking factor permitted by the plant Technical Specifications and at the highest burnup levels anticipated, and assume that the emergency core cooling system (ECCS) is not operating. Use of more realistic assumptions for any of these parameters would increase estimated times to fuel rod failure by factors of two or more. Nevertheless, the use of conservative assumptions in estimating fuel rod failure times is considered appropriate since such failure times are likely to be used primarily in consideration of the necessary closure time for certain containment isolation valves. Since it is important that closure of such valves be ensured before the release of significant radioactivity to the environment, a conservative estimate of fuel failure time and consequent onset of fission product appearance is deemed appropriate. For plants with leak-before-break approval for their reactor coolant system piping, a longer duration before fuel clad failure is expected. However, other constraints may become the limiting factor on containment isolation valve closure time.

4.3 Release Phase Durations

The durations of the various release phases have been selected primarily by examination of the values available for the group of severe accident scenarios considered in Section 3. The durations of the early in-vessel and ex-vessel release phases differs for BWRs versus PWRs and reflect the differing core heatup rates as well as the differing amounts of zirconium available to supply chemical energy after core-melt. While the selected durations of the release phases are realistic, some conservatisms should be noted. The duration of the early in-vessel release phase for BWRs and PWRs is short and does not represent a probabilistically weighted average or mean value for the accident sequences considered. This will introduce a given quantity of fission products into containment in a shorter time than might be expected for a typical sequence.

Similarly, the duration of the ex-vessel release phase, while considered realistic for the bulk of the fission products being released, is short for releases of tellurium and ruthenium since, as noted in Section 3.3, release of these nuclides occurs over a longer time.

The selected release duration times have been chosen primarily on the basis of simplicity, since an accurate determination of the duration of the release phases depends not only on the reactor type but also on the applicable accident sequence, which varies for each reactor design.

4.4 Composition and Magnitude of Releases

The composition of the fission products was initially based on the grouping developed with the STCP, but has been modified as discussed in Section 3.4.

The magnitudes of the fission products released into containment for the accident source term were selected in the draft version of this report to be the mean values, using NUREG-1150 methodology, for BWR and PWR low-pressure scenarios involving high estimates of zirconium oxidation. The uncertainty distributions for the in-vessel release and total release into containment are displayed graphically in Appendix A. Bounding estimates for the releases into containment taken from Reference 17, using the STCP methodology, are shown in Appendix B.

The release magnitudes for the low volatile fission products were reduced significantly in the final report. This reduction was based upon recent experimental research results (Ref. 25) since completion of NUREG-1150, as well as a re-examination of the uncertainty distribution, in response to comments on the draft report. Research on in-vessel phenomena includes the in-pile Severe Fuel Damage (SFD) experiments in the Power Burst Facility (PBF), further

examination of the Three Mile Island (TMI) accident, and the SASCHA out-of-pile tests. Ex-vessel insights derive primarily from large scale tests performed as part of the internationally sponsored ACE Program. Reference 25 notes that, based on the SFD experiments as well as the TMI accident, in-vessel release fractions for cerium, for example, were about 10^{-4}, compared to the value of 10^{-2} cited in the draft report. Based on these results, the NRC staff concludes that the low volatile release fractions cited in draft NUREG-1465 are too high.

The uncertainty distributions were also examined to obtain additional insight. As can be seen from the uncertainty distributions in Appendix A, the range of release estimates for the volatile nuclides, such as the noble gases, iodine, cesium, and to some extent tellurium, spans about one order of magnitude. For this group of nuclides, use of the mean value is a reasonable estimate of the release fraction. In contrast, the range for the low volatile nuclides, such as barium, strontium, cerium and lanthanum, spans about 4 to 6 orders of magnitude. For the latter group of nuclides, the mean value can be misleading, since it may be well in excess of other measures of the distribution. This is illustrated in Table 4.1 which tabulates the mean, median, and 75th percentile values for several low volatile nuclides released during the early in-vessel phase.

Table 4.1 Measures of Low Volatile In-Vessel Release Fractions

Nuclide	Mean	Median	75th percentile
Sr	0.03	0.001	0.006
Ba	0.04	0.003	0.009
La	0.002	0.00003	0.0003
Ce	0.01	0.00006	0.0006

As can be seen from Table 4.1, the mean value for this group of nuclides is one to two orders of magnitude greater than the median value, and is about 5 times greater than the 75th percentile of the distribution. For this group of nuclides, the mean is controlled by the upper tail of the distribution, and the details of the whole distribution may be more indicative of the uncertainty than the "bottom line" results, such as a mean value. Because of this, the final version of this report has chosen not to use the mean value in estimating releases for the non-volatile nuclides. While the median value might be selected as an alternate, it fails to provide an appreciation of the range of values lying above it. Since this report is intended for regulatory applications, the intent is to avoid under-estimation of potential releases or offsite doses, without undue conservatism. Hence, for the final report, the 75th percentile value has been selected for the low volatile nuclides on the basis that it bounds most of the range of values, without undue influence by the upper tail of the distribution.

Uncertainties, particularly in understanding and modeling core melt progression phenomena, can affect the duration of the early in-vessel release phase, including the timing of reactor pressure vessel failure. An increase in duration of the early in-vessel phase can lead to increased releases of volatile fission products during the early in-vessel phase and a concomitant reduction during the ex-vessel phase. An increase in duration of the early in-vessel phase, however, also provides additional time for fission product removal within containment by natural processes or fission product cleanup systems.

Upper bound estimates, tabulated in Appendix B, indicate that virtually all the iodine and cesium could enter the containment. Similarly, for tellurium, upper bound estimates indicate that as much as about two-thirds of the core inventory of tellurium could be released into containment. Hence, for this important group of radionuclides (iodine, cesium, and tellurium), the upper bound estimates of total release into containment are approximately 1.5 times the mean value estimates.

For the lower volatility radionuclides such as barium and strontium, upper bound estimates range from about 50 to 70% of the core inventory released into containment. Almost all of this is estimated to be released as a result of core-concrete interactions. In contrast, mean value estimates range from 15 to 25%. Hence, in this case, the upper bound estimates are about two to three times the mean values.

Finally, for the refractory nuclides such as lanthanum and cerium, the upper bound estimates indicate that about 5% of the inventory of these nuclides could appear within containment, whereas the mean value estimate indicates only about 1% released.

PRAs have indicated that, considering the magnitudes of the radioactive species estimated to be released to the environment for severe reactor accidents, the radionuclides having the greatest impact on risk are typically the volatile nuclides such as iodine and cesium, with tellurium to a somewhat lesser degree. The uncertainty distributions for this group of radionuclides is also the smallest, as shown in the graphical tabulations of Appendix A. Hence, our ability to predict the behavior and releases for this group of nuclides is significantly better than for other fission product groupings.

Mean value estimates selected for the in-containment accident source term provide reasonable estimates for the important nuclides consisting of iodine, cesium, and tellurium. These estimates show a relatively low degree of uncertainty and are unlikely to be exceeded by more than 50%. Uncertainty increases in estimating releases for the remaining nuclides.

4.5 Iodine Chemical Form

The chemical form of iodine entering containment was investigated in Reference 18. On the basis of this work, the NRC staff concludes that iodine entering containment from the reactor coolant system is composed of at least 95% cesium iodide (CsI), with no more than 5% I plus HI. Once within containment, highly soluble cesium iodide will readily dissolve in water pools and plate out on wet surfaces in ionic form. Radiation-induced conversion of the ionic form to elemental iodine will potentially be an important mechanism. If the pH is controlled to a level of 7 or greater, such conversion to elemental iodine will be minimal. If the pH is not controlled, however, a relatively large fraction (greater for PWRs than BWRs) of the iodine dissolved in containment pools in ionic form will be converted to elemental iodine.

5 IN-CONTAINMENT REMOVAL MECHANISMS

Since radioactive fission products within containment are in the form of gases and finely divided airborne particulates (aerosols), the principal mechanism by which fission products find their way from the reactor to the environment with an intact containment is via leakage from the containment atmosphere. The specific fission product inventory present in the containment atmosphere at any time depends on two factors: (1) the source, i.e., the rate at which fission products are being introduced into the containment atmosphere, and (2) the sink, the rate at which they are being removed. Aspects of the release and transport of fission products from the core into the containment atmosphere were presented in Section 3.

Mechanisms that remove fission products from the atmosphere with consequent mitigation of the in-containment source term fall into two classes: (1) engineered safety features (ESFs) and (2) natural processes. ESFs to remove or reduce fission products within the containment are presently required (Criterion 41 in Appendix A of 10 CFR Part 50) and include such systems as containment atmosphere sprays, BWR suppression pools, and filtration systems utilizing both particulate filters and charcoal adsorption beds for the removal of iodine, particularly in elemental form. Natural removal includes such processes as aerosol deposition and the sorption of vapors on equipment and structural surfaces.

The draft version of this report contained a discussion of some of the more important fission product removal mechanisms, including some quantitative results. These numerical results were intended to be illustrative of the phenomena involved and were not intended to be applied rigorously, however. It was recognized that the data and illustrations used in the draft might not be applicable to all situations.

In recognition of this, the NRC staff undertook to examine, with contractor assistance, improved understanding of fission product removal mechanisms. At this time, this effort is still underway. Rather than provide numerical values that may be inapplicable, this report will provide references, where available, so that the reader may utilize improved methodologies to obtain results that apply to the situation at hand.

5.1 Containment Sprays

Containment sprays, covered in Standard Review Plan (SRP) Section 6.5.2 (Ref. 28), are used in many PWR designs to provide post-accident containment cooling as well as to remove released radioactive aerosols. Sprays are effective in reducing the airborne concentration of elemental and particulate iodines as well as other particulates, such as cesium, but are not effective in removing noble gases or organic forms of iodine. The reduction in airborne radioactivity within containment by a spray system as a function of time is expressed as an exponential reduction process, where the spray removal coefficient, lambda, is taken to be constant over a large part of the regime. Typical PWR containment spray systems are capable of rapidly reducing the concentration of airborne activity (by about 2 orders of magnitude within about 30 minutes, where both spray trains are operable). Once the bulk of the activity has been removed, however, the spray becomes significantly less effective in reducing the remaining fission products. This is usually accounted for by either employing a spray cut-off, wherein the spray removal becomes zero after some reduction has been achieved, or changing to a much smaller value of lambda to reflect the decreased removal effectiveness of the spray when airborne concentrations are low.

SRP Section 6.5.2 (Ref. 28) provides expressions for calculating spray lambdas, depending on plant parameters as well as the type of species removed. In addition, SRP 6.5.2 currently suggests that the containment sump solution be maintained at values at or above pH levels of 7, commencing with spray recirculation, to minimize revolatilization of iodine in the sump water. Current guidance states that

containment spray systems be initiated automatically, because of the instantaneous appearance of the source term within containment, and that the spray duration not be less than 2 hours. In contrast, the revised source term information given in Section 3 suggests that spray system actuation might be somewhat delayed for radiological purposes, but that the spray system duration should be for a longer period of about 10 or more hours. Because sprays are effective in rapidly removing particulates from the containment atmosphere, intermittent operation over a prolonged period may also provide satisfactory mitigation.

The spray removal coefficient for particulates appears particularly important in view of the information presented in Section 3, which indicates that most fission products are expected to be in particulate form. The spray removal coefficient (λ) is derived from the following equation from Standard Review Plan Section 6.5.2

$$\lambda = \frac{3hFE}{2VD}$$

h = Fall height of spray drops
V = Containment building volume
F = Spray flow
E/D = the ratio of a dimensionless collection efficiency E to the average spray drop Diameter D. E/D is conservatively assumed to be equal to 10/meter for spray drops 1 mm in diameter changing to 1/meter when the aerosol mass has been depleted by a factor of 50.

Using values typical for PWRs, the formulation given in SRP 6.5.2 estimates particulate removal rates to be on the order of 5 per hour. Nourbakhsh (Ref. 29) examined the effectiveness of containment sprays, as evaluated in NUREG-1150 (Ref. 7), in decontaminating both in-vessel and ex-vessel releases. Powers and Burson (Ref. 30) have developed a more realistic, yet simplified, model with regard to evaluating the effectiveness of aerosol removal by containment sprays

5.2 BWR Suppression Pools

BWRs use pressure suppression pools to condense steam resulting from a loss-of-coolant accident. Prior to the release to the reactor building, these pools also scrub radioactive fission products that accompany the steam. Regulatory Guide 1.3 (Ref. 2) suggests not allowing credit for fission product scrubbing by BWR suppression pools, but SRP Section 6.5.5 (Ref. 31) was revised to suggest allowing such credit. The pool water will retain soluble, gaseous, and solid fission products such as iodines and cesium but provide no attenuation of the noble gases released from the core. The Reactor Safety Study (WASH-1400, Ref. 5) assumed a decontamination factor (DF) of 100 for subcooled suppression pools and 1.0 for steam saturated pools. Since 1975 when WASH-1400 was published, several detailed models have been developed for the removal of radioactive aerosols during steam flow through suppression pools.

Calculations for a BWR with a Mark I containment (Ref. 27) used in NUREG-1150 (Ref. 7) indicate that DFs ranged from 1.2 to about 4000 with a median value of about 80. The suppression pool has been shown to be effective in scrubbing some of the most important radionuclides such as iodine, cesium, and tellurium, as these are released in the early in-vessel phase. The NRC staff is also presently reviewing fission product scrubbing by suppression pools to develop simplified models.

If not bypassed, the suppression pool will also be effective in scrubbing ex-vessel releases. Suppression pool bypass is an important aspect that places an upper limit on the overall performance of the suppression pool in scrubbing fission products. For example, if as little as 1% of the fission products bypass the suppression pool, the effective DF, taking bypass into account, will be less than 100, regardless of the pool's ability to scrub fission products.

Although decontamination factors for the suppression pool are significant, the potential for iodine re-evolution can be important. Re-evolution of iodine was judged to be important in accident sequences where the containment had failed and the suppression pool was boiling. There is presently no requirement for pH control in BWR suppression pools. Hence, it is possible that suppression pools would scrub substantial amounts of iodine in the early phases of an accident, only to re-evolve it later as elemental iodine. It may well be that additional materials likely to be in the suppression pool as a result of a severe accident, such as cesium borate or cesium hydroxide and core-concrete decomposition products, would counteract any reduction in pH from radiolysis and would ensure that the pH level was sufficiently high to preclude re-evolution of elemental iodine. Therefore, if credit is to be given for long-term retention of iodine in the suppression pool, maintenance of the pH at or above a level of 7 must be demonstrated. It is important to note, however, that this is not a matter of concern for present plants since all BWRs employ safety-related filtration systems (see Section 5.3) designed to cope with large quantities of elemental iodine. Hence, even if the suppression pool were to re-evolve significant amounts of elemental iodine, it would be retained by the existing downstream filtration system.

5.3 Filtration Systems

ESF filtration systems are discussed in Regulatory Guide 1.52 (Ref. 32) and are used to reduce the radioactive aerosols and iodine released during postulated accident conditions.

A typical ESF filtration system consists of redundant trains that each have demisters to remove steam and water droplets from the air entering the filter bank, heaters to reduce the relative humidity of the air, high efficiency particulate air (HEPA) filters to remove particulates, charcoal adsorbers to remove iodine in elemental and organic form, followed finally by additional HEPA filters to remove any charcoal fines released.

Charcoal adsorber beds can be designed, as indicated in Regulatory Guide 1.52, to remove from 90 to 99% of the elemental iodine and from 30 to 99% of the organic iodide, depending upon the specific filter train design.

Revised insights on accident source terms, given in Section 3, may have several implications for ESF filtration systems. Present ESF filtration systems are not sized to handle the mass loadings of non-radioactive aerosols that might be released as a result of the ex-vessel release phase, which could produce releases of significant quantities of nonradioactive as well as radioactive aerosols. However, if ESF filtration systems are employed in conjunction with BWR suppression pools or if significant quantities of water are overlaying molten core debris (see Section 5.4), large quantities of nonradioactive (as well as radioactive) aerosols will be scrubbed and retained by these water sources, thereby reducing the aerosol mass loads upon the filter system.

A second implication of revised source term insights for ESF filtration systems is the impact of revised understanding of the chemical form of iodine within containment. Present ESF filtration systems presume that the chemical form of iodine is primarily elemental iodine, and these systems include charcoal adsorber beds to trap and retain elemental iodine. Assuming that pH control is maintained within the containment, a key question is whether charcoal beds are necessary. Two questions appear to have a bearing on this issue and must be addressed, even assuming pH control. These are (1) to what degree will CsI retained on particulate filters decompose to evolve elemental iodine? and (2) what effect would hydrogen burns have on the chemical form of the iodine within containment? Based on preliminary information, CsI retained on particulate filters as an aerosol appears to be chemically stable provided that it is not exposed to moisture. Exposure to moisture, however, would lead to CsI decomposition and production of iodine in ionic form (I⁻), which in turn would lead to re-evolution of elemental iodine. Although ESF filtration systems are equipped with demisters and heaters to remove significant moisture before it reaches the charcoal adsorber bed, an additional concern is that the demisters themselves may trap some CsI aerosol.

In conclusion, present ESF filtration systems, while optimized to remove iodine, particularly in elemental form, have HEPA filters that are effective in the removal of particulates as well. Although such filtration systems are not designed to handle the large mass loadings expected as a result of ex-vessel releases, when they are used in conjunction with large water sources such as BWR suppression pools or significant water depths overlaying core debris, the water sources will reduce the aerosol mass loading on the filter system significantly, making such filter systems effective in mitigation of a large spectrum of accident sequences.

5.4 Water Overlying Core Debris

Experimental measurements (Ref. 33) have shown that significant depths of water overlying any molten core debris after reactor pressure vessel failure will scrub and retain particulate fission products. The question of coolability of the molten debris as a result of water overlying it is still under investigation. A major factor that may affect the degree of scrubbing is whether the water layer in contact with the molten debris is boiling or not.

Results from Ref. 33 indicate that both subcooled as well as boiling water layers having a depth of about 3 meters had measured DFs of about 10. A recent study (Ref. 34) performed for the NRC has provided a simplified model to determine the degree of aerosol scrubbing by a water pool overlying core debris interacting with concrete.

5.5 Aerosol Deposition

Since the principal pathway for transport of fission products is via airborne particulates, i.e., aerosols, this subject is discussed in some detail. Aerosols are usually thought of as solid particulates, but in general, the term also includes finely divided liquid droplets such as water, i.e., fog. The two major sources of aerosols are condensation and entrainment. Condensation aerosols form when a vapor originating from some high-temperature source moves into a cooler region where the vapor falls below its saturation temperature and nucleation begins. Entrainment aerosols form when gas bubbles break through a liquid surface and drag droplets of the liquid phase into the wake of the bubble as it leaves the surface. In general, condensation particles are smaller in size (submicron to a few microns), while entrainment particles are usually larger (1.0–100 microns). Once airborne, both types of aerosols behave in a similar manner with respect to both natural and engineered removal processes.

There are four natural processes that remove aerosols from the containment atmosphere over a period of time: (1) gravitational settling, (2) diffusiophoresis, (3) thermophoresis, and (4) particle diffusion. (Particle diffusion is less important than the first three processes and will not be discussed further.) All particles fall naturally under the force of gravity and collect on any available surface that terminates the fall, e.g., the floor or upper surfaces of equipment. Both diffusiophoresis and thermophoresis cause the deposition of aerosol particles on all surfaces regardless of their orientation, i.e., walls and ceiling as well as the floor. Diffusiophoresis is the process by which water vapor in the atmosphere 'drags' aerosol particles with it as it migrates (diffuses) toward a relatively cold surface on which condensation is taking place. Thermophoresis also causes aerosol particles to move toward and deposit on colder surfaces but not as a result of mass motion. Rather, the decreasing average velocity of the surrounding gas molecules tends to drive the particle down the temperature gradient until it traverses the interface layer and comes into contact with the surface where it sticks.

Aerosol agglomeration is another natural phenomenon that has an influence on the rates at which the removal processes described above will proceed. Agglomeration results from the random inelastic collisions of particles with each other. The process brings about a gradual increase in average particle size resulting in more rapid gravitational settling. Three phenomena contribute to particle growth by agglomeration: (1) Brownian motion, (2) gravitational fall, and (3) turbulence. Brownian agglomeration is caused by particle collisions resulting from random 'buffeting' by high-energy gas molecules. Gravitational agglomeration results from the fact that some particles fall faster than others and therefore tend to collide with and stick to other slower falling particles on their way down. Finally, rapid variations in gas velocity and flow direction in the atmosphere, i.e., turbulence, tend to increase the rate at which particle collisions occur and thus increase the average particle size. It is to be expected that, as agglomeration advances, the size of the particle will increase, and its shape can be expected to change as well. These latter factors have a strong influence on the removal processes.

The agglomeration and aerosol removal processes all depend critically upon the thermodynamic state and thermal-hydraulic conditions of the containment atmosphere. For example, the condensation onto and evaporation of water from the aerosol particles themselves have strong effects on all of the agglomeration and removal processes. Water condensed on aerosol particles increases their mass and makes them more spherical; both of these effects tend to increase the rate of gravitational settling. Some aerosols, such as CsI and CsOH, are hygroscopic and absorb water vapor even when the containment atmosphere is below saturation. As with condensation, hygroscopicity also increases the rate of deposition.

Because of its importance to fields such as weather and atmosphere pollution, the behavior of aerosols has been under study for many decades. A number of computer codes have been developed to specifically consider aerosol behavior as it relates to nuclear accident conditions. The most complete mechanistic treatment of aerosol behavior in the reactor containment is found in CONTAIN, a computer code developed at Sandia National Laboratories under NRC sponsorship for the analysis of containment behavior under severe accident conditions. The aerosol models in the NAUA code are very similar to those used in CONTAIN; NAUA was developed at the Kernforschungszentrum, Karlsruhe, F.R.G., and was used for aerosol treatment in the NRC STCP. There are a number of other well-known aerosol behavior computer codes, but these two are the most widely used and accepted throughout the international nuclear safety community.

The rate at which gravitational settling occurs depends upon the degree of agglomeration at any particular time (i.e., the average particle size) as well as the total particle density m (mass per unit volume). Thus, as in most cases where the decrement of a variable is proportional to the variable itself, one can expect an exponential behavior. The gravitational settling process is quite complex and depends upon a large number of physical quantities, e.g., collision shape factor, particle settling shape factor, gas viscosity, effective settling height, density correction factor, normalized Brownian collision coefficient, gravitational acceleration, and particle material density. The only variable in this list that is independent of the plant, the accident scenario, and the atmospheric thermal-hydraulic conditions is the constant of gravitation. It follows that no single DF can be ascribed to cover the entire range of plant designs, accident scenarios, and source materials. An effort is under way to establish a set of simplified algorithms that can be used to provide a set of specific ranges of atmosphere conditions. This effort is still underway at this time.

6. REFERENCES

1. U.S. Nuclear Regulatory Commission; "Reactor Site Criteria," Title 10, Code of Federal Regulations (CFR), Part 100.

2. U.S. Nuclear Regulatory Commission; "Assumptions Used for Evaluating the Potential Radiological Consequences of a Loss of Coolant Accident for Boiling Water Reactors," Regulatory Guide 1.3, Revision 2, June 1974.

3. U.S. Nuclear Regulatory Commission; "Assumptions Used for Evaluating the Potential Radiological Consequences of a Loss of Coolant Accident for Pressurized Water Reactors," Regulatory Guide 1.4, Revision 2, June 1974.

4. J.J. DiNunno et al., "Calculation of Distance Factors for Power and Test Reactor Sites," Technical Information Document (TID)-14844, U.S. Atomic Energy Commission, 1962.

5. U.S. Nuclear Regulatory Commission; "Reactor Safety Study: An Assessment of Accident Risks in U.S. Commercial Nuclear Power Plants," WASH-1400 (NUREG-75/014), December 1975.

6. J. A. Gieseke et al., "Source Term Code Package: A User's Guide," NUREG/CR-4587 (BMI-2138), prepared for NRC by Battelle Memorial Institute, July 1986.

7. U.S. Nuclear Regulatory Commission; "Severe Accident Risks: An Assessment for Five U.S. Nuclear Power Plants," NUREG-1150, December 1990.

8. M.R. Kuhlman, D.J. Lehmicke, and R.O. Meyer, "CORSOR User's Manual," NUREG/CR-4173 (BMI-2122), prepared for NRC by Battelle Memorial Laboratory, March 1985.

9. H. Jordan, and M.R. Kuhlman, "TRAP-MELT2 User's Manual," NUREG/CR-4205 (BMI-2124), prepared for NRC by Battelle Memorial Laboratory, May 1985.

10. D.A. Powers, J.E. Brockmann, and A.W. Shiver, "VANESA: A Mechanistic Model of Radionuclide Release and Aerosol Generation During Core Debris Interactions with Concrete," NUREG/CR-4308 (SAND 85-1370), prepared for NRC by Sandia National Laboratories, July 1986.

11. P.C. Owczarski, A.K. Postma, and R.I. Schreck, "Technical Bases and User's Manual for the Prototype of SPARC—A Suppression Pool Aerosol Removal Code," NUREG/CR-3317 (PNL-4742), prepared for NRC by Battelle Pacific Northwest Laboratories, May 1985.

12. W.K. Winegardner, A.K. Postma, and M.W. Jankowski, "Studies of Fission Product Scrubbing within Ice Compartments," NUREG/CR-3248 (PNL-4691), prepared for NRC by Battelle Pacific Northwest Laboratories, May 1983.

13. H. Bunz, M. Kayro, and W. Schock, "NAUA-Mod 4: A Code for Calculating Aerosol Behavior in LWR Core Melt Accidents," KfK-3554, Kernforschungszentrum Karlsruhe Germany, 1983.

14. R.M. Summers, et al., "MELCOR 1.8.0: A Computer Code for Nuclear Reactor Severe Accident Source Term and Risk Assessment Analysis," NUREG/CR-5531 (SAND 90-0364), prepared for NRC by Sandia National Laboratories, January 1991.

15. K.R. Jones, et al, "Timing Analysis of PWR Fuel Pin Failures," NUREG/CR-5787 (EGG-2657), prepared for NRC by Idaho National Engineering Laboratory, September 1992.

16. H.P. Nourbakhsh, M. Khatib-Rahbar, and R.E. Davis, "Fission Product Release Characteristics into Containment Under Design Basis and Severe Accident Conditions," NUREG/CR-4881 (BNL-NUREG-52059), prepared for NRC by Brookhaven National Laboratory, March 1988.

17. H.P. Nourbakhsh,: "Estimates of Radionuclide Release Characteristics into Containment Under Severe Accidents," NUREG/CR-5747 (BNL-NUREG-52289), prepared for NRC by Brookhaven National Laboratory, November 1993.

18. E.C. Beahm, C.F. Weber, and T.S. Kress, "Iodine Chemical Forms in LWR Severe Accidents", NUREG/CR-5732 (ORNL/TM-11861), prepared for NRC by Oak Ridge National Laboratory, April 1992.

19. J.J. Carbajo, "Severe Accident Source Term Characteristics for Selected Peach Bottom Sequences Predicted by the MELCOR Code," NUREG/CR-5942 (ORNL/TM-12229), prepared for NRC by Oak Ridge National Laboratory, September 1993.

20. D.J. Alpert, D.I. Chanin, and L.T. Ritchie, "Relative Importance of Individual Elements to Reactor Accident Consequences Assuming Equal Release Fractions."NUREG/CR-4467, prepared for NRC by Sandia National Laboratories, 1986.

21. C.F. Weber, E.C. Beahm and T.S. Kress, "Models of Iodine Behavior in Reactor Containments," ORNL/TM-12202, Oak Ridge National Laboratory, October 1992.

22. E.C. Beahm, R.A. Lorenz, and C.F. Weber, "Iodine Evolution and pH Control," NUREG/CR-5950, (ORNL/TM-12242), prepared for NRC by Oak Ridge National Laboratory, December 1992.

23. A.K. Postma, and R.W. Zavadowski, "Review of Organic Iodide Formation Under Accident Conditions in Water-Cooled Reactors," WASH-1233, U.S. Atomic Energy Commission, October 1972.

24. E.C. Beahm, W.E. Shockley, and O.L. Culberson, "Organic Iodide Formation Following Nuclear Reactor Accidents," NUREG/CR-4327, (ORNL/TM-9627), prepared for NRC by Oak Ridge National Laboratory, December 1985.

25. D. J. Osetek, "Low Volatile Fission Product Releases During Severe Reactor Accidents," DOE/ID-13177-2, prepared for U.S. Department of Energy by Los Alamos Technical Associates, October 1992.

26. M. Silberberg et al., "Reassessment of the Technical Bases for Estimating Source Terms," NUREG-0956, July 1986.

27. R.S. Denning, et al., "Radionuclide Release Calculations for Selected Severe Accident Scenarios: BWR Mark I Design," NUREG/CR-4624, Vol. 1, prepared for NRC by Battelle Memorial Institute, July 1986.

28. U.S. Nuclear Regulatory Commission: "Containment Spray as a Fission Product Cleanup System," Standard Review Plan, Section 6.5.2, Revision 2, NUREG-0800, December 1988.

29. H.P. Nourbakhsh,: "In-Containment Removal Mechanisms," Presentation to NRC staff January 3, 1992, Brookhaven National Laboratory, January 1992.

30. D.A. Powers and S.B. Burson, "A Simplified Model of Aerosol Removal by Containment Sprays," NUREG/CR-5966, (SAND92-2689), prepared for NRC by Sandia National Laboratories, June 1993.

31. U.S. Nuclear Regulatory Commission: "Pressure Suppression Pool as a Fission Product Cleanup System," Standard Review Plan, Section 6.5.5, NUREG-0800, December 1988.

32. U.S. Nuclear Regulatory Commission: "Design, Testing, and Maintenance Criteria for Postaccident Engineered-Safety-Feature Atmosphere Cleanup System Air Filtration and Adsorption Units of Light-Water-Cooled Nuclear Power Plants," Regulatory Guide 1.52, Revision 2, March 1978.

33. J. Hakii et al., "Experimental Study on Aerosol Removal Efficiency for Pool Scrubbing Under High Temperature Steam Atmosphere," Proceedings of the 21st DOE/NRC Nuclear Air Cleaning Conference, August 1990.

34. D.A. Powers and J.L. Sprung, "A Simplified Model of Aerosol Scrubbing by a Water Pool Overlying Core Debris Interacting With Concrete," NUREG/CR-5901, (SAND92-1422), prepared for NRC by Sandia National Laboratories, November 1993.

APPENDIX A
UNCERTAINTY DISTRIBUTIONS

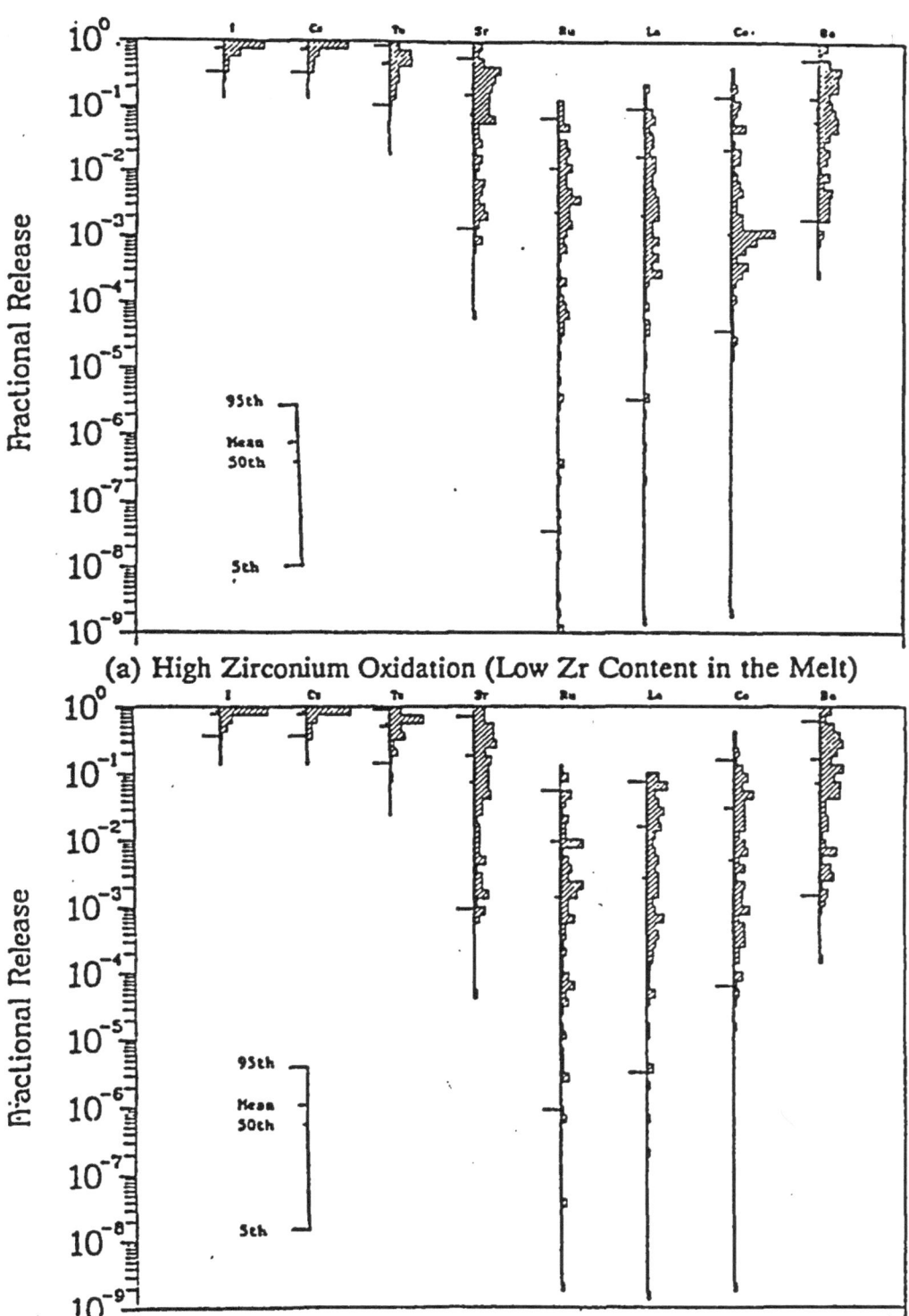

(a) High Zirconium Oxidation (Low Zr Content in the Melt)

(b) Low Zirconium Oxidation (High Zr Content in the Melt)

Uncertainty Distributions for Total Releases Into Containment PWR, Low RCS Pressure, Limestone Concrete, Dry Cavity, Two Openings After VB, FPART = 1.

NUREG-1465

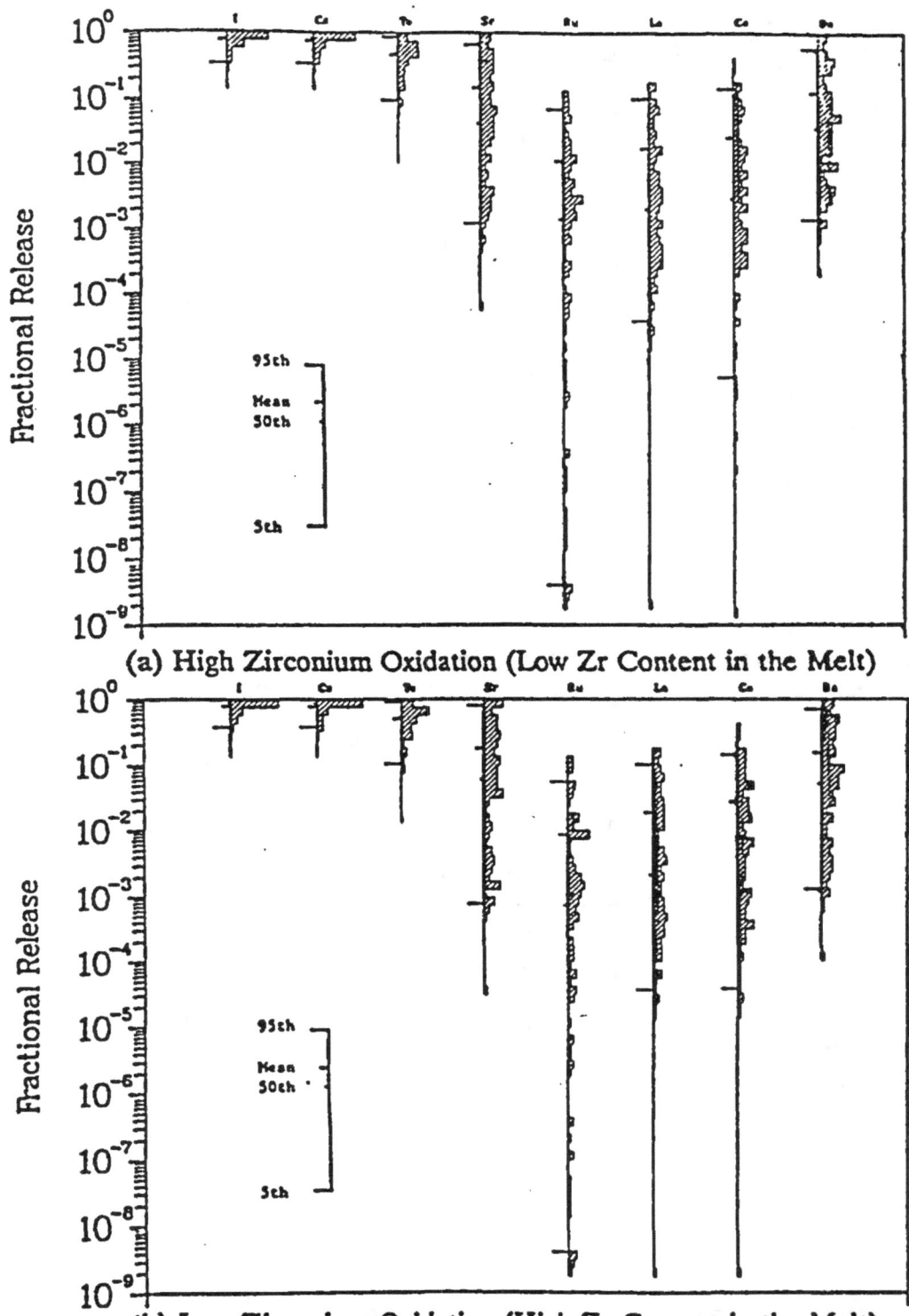

(a) High Zirconium Oxidation (Low Zr Content in the Melt)

(b) Low Zirconium Oxidation (High Zr Content in the Melt)

Uncertainty Distributions for Total Releases Into Containment PWR, Low RCS Pressure, Basaltic Concrete, Dry Cavity, Two Openings After VB, FPART = 1.

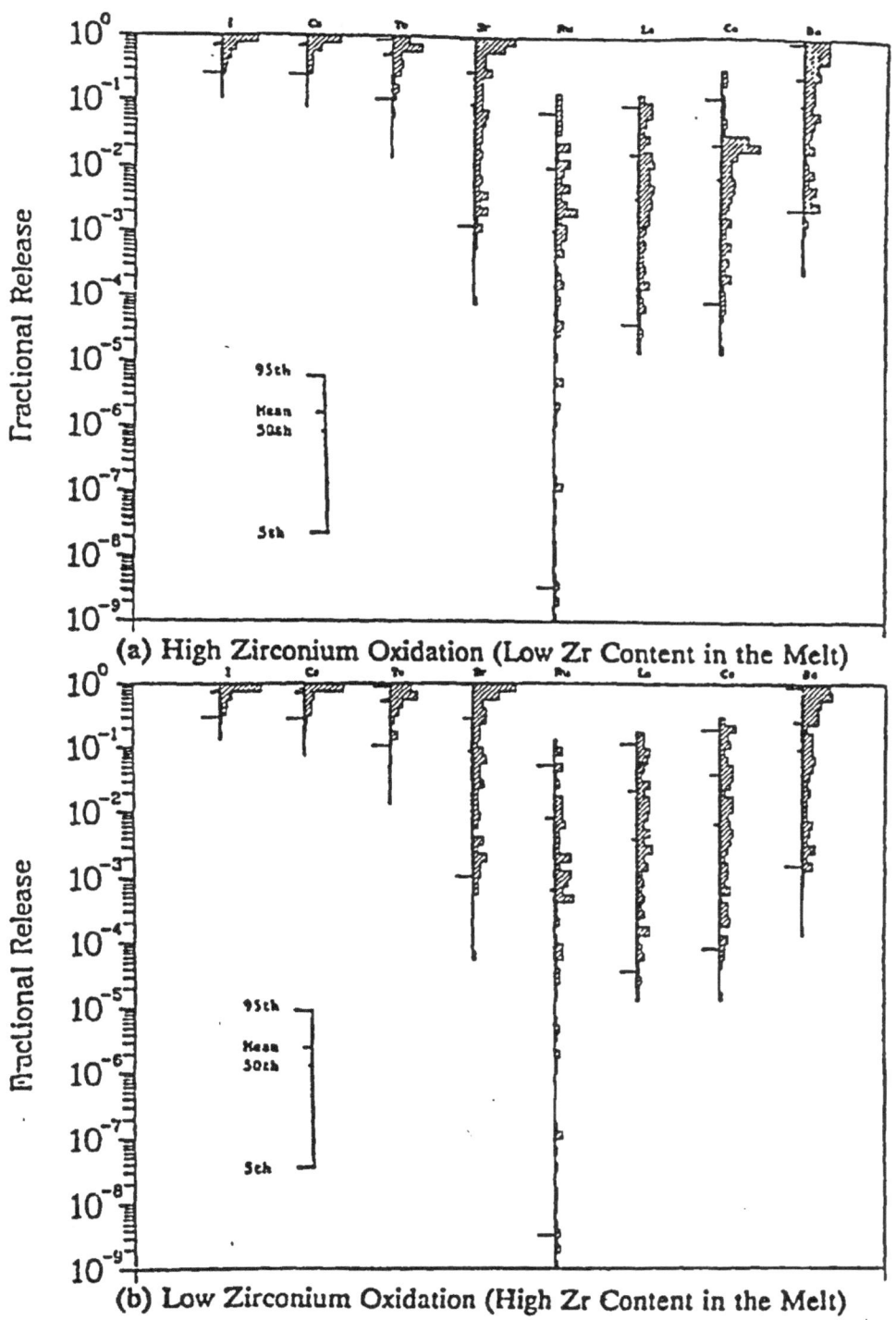

(a) High Zirconium Oxidation (Low Zr Content in the Melt)

(b) Low Zirconium Oxidation (High Zr Content in the Melt)

Uncertainty Distributions for Total Releases Into Containment BWR, Low Pressure Fast Station Blackout, Limestone Concrete, Dry Pedestal, Low Drywell Temperature, FPART = 1.

27

APPENDIX B

STCP BOUNDING VALUE RELEASES

Updated Bounding Value of Radionuclide Releases Into the Containment Under Severe Accident Conditions for PWRs

| | ST$_{INV}$ [a] | | ST$_{VB}$ | ST$_{EXV}$ [e] | | ST$_{REV}$ | |
	High RCS Pressure	Low RCS Pressure	High RCS Pressure	Limestone Concrete	Basaltic Concrete	High RCS Pressure	Low RCS Pressure
NG	1.0	1.0	0.	0.	0.	0.	0.
I	0.30	0.75	0.10	0.15	0.15	0.05	0.02
Cs	0.30	0.75	0.10	0.15	0.15	0.02	0.02
Te	0.20	0.50	0.05	0.40	0.30	0.02	0.01
Sr-Ba	0.003	0.01	0.01	0.40	0.15	–	–
Ru	0.003	0.01	0.05	0.005	0.005	–	–
La-Ce	5×10^{-5}	1.5×10^{-4}	0.01	0.05	0.05	–	–
Release Duration	40 minutes			2 hours [c]		10 hours	

(a) All entries are fractions of initial core inventory.

(b) Assuming 100% of the core participate in CCI.

(c) Except for Te and Ru where the duration is extended to five hours.

29   NUREG-1465

Updated Bounding Value of Radionuclide Releases Into the Containment Under Severe Accident Conditions for BWRs

| | ST_{INV}[a] | | ST_{VB} | ST_{ENV}[c] | | ST_{REV} | |
	High RCS Pressure	Low RCS Pressure[b]	High RCS Pressure	Limestone Concrete	Basaltic Concrete	High RCS Pressure	Low RCS Pressure[b]
NG	1.	1.	0.	0.	0.	0.	0.
I	0.50	0.75	0.10	0.15	0.15	0.10	0.02
Cs	0.50	0.75	0.10	0.15	0.15	0.05	0.01
Te	0.10	0.15	0.05	0.50	0.30	0.02	0.02
Sr-Ba	0.003	0.01	0.01	0.70	0.30	--	--
Ru	0.003	0.01	0.05	0.005	0.005	--	--
La-Ce	5×10^{-3}	1.5×10^{-4}	0.01	0.10	0.10	--	--
Release Duration	1.5 hours		3 hours[d]			10 hours	

[a] All entries are fractions of initial core inventory.

[b] High pressure ATWS are also considered in this category.

[c] Assuming 100% of the core participate in CCI.

[d] Except for Te and Ru where the duration is extended to six hours.

NRC FORM 335
(2-89)
NRCM 1102,
3201, 3202

U.S. NUCLEAR REGULATORY COMMISSION

BIBLIOGRAPHIC DATA SHEET

(See instructions on the reverse)

1. REPORT NUMBER (Assigned by NRC. Add Vol., Supp., Rev., and Addendum Numbers, if any.)
NUREG-1465

2. TITLE AND SUBTITLE

Accident Source Terms for Light-Water Nuclear Power Plants

3. DATE REPORT PUBLISHED	
MONTH	YEAR
February	1995

4. FIN OR GRANT NUMBER

5. AUTHOR(S)

L. Soffer, S. B. Burson, C. M. Ferrell, R. Y. Lee, J. N. Ridgely

6. TYPE OF REPORT

7. PERIOD COVERED *(Inclusive Dates)*

8. PERFORMING ORGANIZATION – NAME AND ADDRESS *(If NRC, provide Division, Office or Region, U.S. Nuclear Regulatory Commission, and mailing address; if contractor, provide name and mailing address.)*

Division of Systems Technology
Office of Nuclear Regulatory Research
U.S. Nuclear Regulatory Commission
Washington, DC 20555-0001

9. SPONSORING ORGANIZATION – NAME AND ADDRESS *(If NRC, type "Same as above"; if contractor, provide NRC Division, Office or Region, U.S. Nuclear Regulatory Commission, and mailing address.)*

Same as above

10. SUPPLEMENTARY NOTES

11. ABSTRACT *(200 words or less)*

In 1962 the U.S. Atomic Energy Commission published TID-14844, "Calculation of Distance Factors for Power and Test Reactors" which specified a release of fission products from the core to the reactor containment for a postulated accident involving "substantial meltdown of the core". This "source term", the basis for the NRC's Regulatory Guides 1.3 and 1.4, has been used to determine compliance with the NRC's reactor site criteria, 10 CFR Part 100, and to evaluate other important plant performance requirements. During the past 30 years substantial additional information on fission product releases has been developed based on significant severe accident research. This document utilizes this research by providing more realistic estimates of the "source term" release into containment, in terms of timing, nuclide types, quantities and chemical form, given a severe core-melt accident. This revised "source term" is to be applied to the design of future light water reactors (LWRs). Current LWR licensees may voluntarily propose applications based upon it.

12. KEY WORDS/DESCRIPTORS *(List words or phrases that will assist researchers in locating the report.)*

Severe Accident Source Term
Core Meltdown
Design Basis Accident
TID-14844 Replacement
Core Fission Product Releases

13. AVAILABILITY STATEMENT
Unlimited

14. SECURITY CLASSIFICATION

(This Page)
Unclassified

(This Report)
Unclassified

15. NUMBER OF PAGES

16. PRICE